Mons Meg
a symbol of Scotland
Peter Lead

Mons Meg on her 'new carriage' 1934.

© Peter Lead, 2021
First published in the United Kingdom, 2021,
by Stenlake Publishing Ltd.
www.stenlake.co.uk
ISBN 978-1-84033-920-8

The publishers regret that they cannot supply
copies of any pictures featured in this book.

Printed by
Blissetts, Unit E1-E8 Shield Drive,
West Cross Ind Pk, Brentford, TW8 9EX

Perhaps the most unexpected image to emerge from my researches into the history and associations of *Mons Meg*. A group of soldiers from Hitler's *Kriegsmarine* (navy) posing with a solitary Highland soldier from the period 1934-39.

Contents

Scottish National Symbols 5
The Galloway Legend 11
Early Artillery in Scotland 15
"Fiery Face's" Passion 19
The Duke's Gift 21
Last Siege of Roxburgh 25
Serving James IV 29
Royal Salutes and Exile 31
Home Again 37
Admiring Queens 41
Images and Souvenirs 45

Introduction

Mons Meg is the most famous piece of medieval artillery that has survived into present times. Apart from an enforced sojourn in London, she has been domiciled in Edinburgh Castle since arriving in Scotland in 1457. Closely connected with numerous royal figures from when she was gifted to King James II of Scotland, *Mons Meg* quickly became a symbol of Scottish military strength and national identity. During her exile in London, Scots lamented that "Scotland would never be Scotland again until *Mons Meg* came hame." She is, officially part of the Royal Armouries collection on loan to Historic Scotland. Hopefully, a full restoration to the people of Scotland may happen in the near future.

Peter Lead, Wanlockhead, 2021.

American naval officers from the *USS Arkansas* admiring the view by *Mons Meg*. Taken in 1930 during one of the battleship's goodwill visits to Edinburgh. Notice the cine camera by the officer on the right.

Scottish National Symbols

Scotland's national identity has been shaped by its powerful history although in earlier times this mixed a great deal of myth with reality. Across the world, people recognise many of the cultural icons of the country including the thistle, tartan and whisky to name a few, but there is also a grouping of three treasures each closely representing sovereign power and nationhood. These are the Stone of Scone, the Honours of Scotland and symbolising military strength – the bombard known as *Mons Meg*. All of these national treasures now have a home in Edinburgh Castle.

The Stone of Destiny was, according to legend, the Biblical Jacob's Pillar, the rock on which he laid his head as he saw the ladder leading to heaven. In this elaborate story, an Egyptian Princess took the stone to Spain, and then it found its way to Ireland before landing on the west coast of Scotland. The transfer to Scone was said to have been made on the orders of King Kenneth MacAlpin who traditionally brought together the kingdoms of the Scots and Picts around 843 A.D. This wonderfully creative story underpins ancient Scottish royal authority with a strong Biblical reference that added to the authority of the early kings. A modern geological study has readily identified the stone as being a piece of Perthshire red sandstone but rather than reducing the stone's national significance, it suggests a likely origin as an ancient Pictish royal stone.

Edward removed the stone to Westminster Abbey in 1296, where it was housed in a purpose-built wooden chair called King Edward's Chair. This was a visual confirmation of his claim to be Lord Paramount of Scotland with overarching power. Most English and then British monarchs have been crowned on this famous throne, the last being Queen Elizabeth II in 1953. The treaty signed in 1328, which ended the First Scottish War of Independence, contained a promise to return the stone but this was never honoured.

On Christmas Day 1950 four Scottish students took matters into their own hands and removed the Stone from the Coronation Chair after breaking into Westminster Abbey. The police were unable to trace the Stone until it was placed in Arbroath Abbey by the perpetrators on 11th April 1951. Almost immediately it was returned to London, but a year elapsed before it went back on display in its original setting.

King Edward's Chair with the original Stone of Destiny, in situ.

Left: The Stone of Destiny with James Wishart, custodian of Arbroath Abbey, April 1951. The Scottish student's who 'returned it' to Scotland wrapped it in a Saltire and left it in the Abbey. The Declaration of Arbroath (1320) was a letter to the Pope asking him to recognise Scotland's independence and king.

Below: A print from around 1870, showing James III in coronation robes and wearing a crown all of which may be rather speculative. James was crowned at Kelso Abbey in 1460 at the age of eight and proved to be an unpopular and ineffective monarch. He was killed fighting his disaffected nobles at the battle of Sauchieburn (near Stirling) in 1488.

The year 1996 marked the 700th anniversary of the Auld Alliance between France and Scotland and Edward I's invasion of Scotland and the taking of the Stone. On 3rd July of that year, John Major, the then Prime Minister made a statement about the Stone announcing that it was to be returned to Scotland. The following year, on St. Andrew's Day amid elaborate ceremony, the Stone of Destiny was taken to its new resting place in Edinburgh Castle alongside the Honours of Scotland. The Moderator of the General Assembly of the Church of Scotland observed that:

> the ideal of Scottish nationhood and the reality of Scottish identity have never been wholly obliterated from the hearts of the people. The recovery of this ancient symbol of the Stone cannot but strengthen the proud distinctiveness of the people of Scotland.

A writer in the *Edinburgh Evening Courant* on the return of *Mons Meg* in 1828 wrote something very similar when he recorded the 'triumphal procession to welcome its return to Scotland, as connected with the glory and independence of the country.' Both national treasures had found their way back to Scotland with the involvement and approval of the then reigning monarchs, namely George IV and Elizabeth II, reversing the process started by Edward I so many years before.

When Edward I removed the Stone, he also took the Crown, Sceptre, Sword and Ring from the ill-fated John Balliol. Robert the Bruce was crowned with a hastily-made crown at Scone on 25th March 1306 which fell into English hands following his defeat at Methven in the same year. The Crown, Sceptre and Sword of State were the visible symbols of Scotland's regained independence, but little is known about their form before 1488. During the reigns of James IV and V, the current Honours replaced those used before and these are the ones to be seen in the Crown Room at Edinburgh Castle.

Tradition states that the silver-gilt Sceptre was a gift from Pope Alexander VI in 1494. It is certainly the oldest of the Honours. James IV ordered a Sword of Honour from an Edinburgh cutler in 1502, which was replaced by another Papal gift of the magnificent Sword of State in 1507. It is not known exactly when the Crown of Scotland was made, but it appears in its pre-1540 form in a portrait of James IV painted in 1503. Arches were added to the Crown on the orders of James V in 1532, to make it an imperial crown symbolising his inherited claim to be free of any allegiance to a foreign monarch. In 1540 the base was melted down and recast with the addition of 22 gemstones to the original 20 and a further 41ozs of gold. The new gold came from Crawford Muir, most likely from the area centred around Wanlockhead.

James VI had a second coronation in July 1603, when he united the Crowns of England and Scotland, sitting on the Coronation Chair (and the Stone of Destiny) in Westminster Abbey. Scotland had lost its resident king but the Honours became an even more powerful symbolic substitute. After the execution of Charles I and the defeat of the Scottish forces at Dunbar, the Honours were moved north before Cromwell occupied Edinburgh Castle late in 1650. They appeared for Charles II's Coronation at Scone on 1st January 1651 and at the Parliament held in Perth in June. Then they were despatched to the Castle of Dunnottar with the Cromwellian forces in hot pursuit. Dunnottar was besieged for eight long months and there was a baseless claim that *Mons Meg* was there as well to protect the Honours. As the siege drew to its inevitable conclusion, the Honours were smuggled out of the castle and concealed beneath the floor of the nearby Kirk of Kinneff. The Restoration of Charles II in 1660, saw their return to the safety of Edinburgh Castle when the cannons including *Mons Meg* were fired in salute.

Dunnottar Castle where the Honours of Scotland were housed for safety in the mid-seventeenth-century.

As the reigning monarchs rarely visited Scotland, the Honours reverted to their alternate role as symbols of an absent king. At all meetings of Parliament between 1661 and 1707, they were taken with elaborate ceremony and procession from Edinburgh Castle to the Parliament House, on the Royal Mile. This was also an occasion when all the cannons in the Castle batteries were always discharged in salute. The Act of Union in 1707 brought the kingdoms of England and Scotland together and in essence the Honours became redundant, but one clause of the Articles of Union ensured that they would not be removed to London stating that: '*the Crown, Sceptre and Sword of State ... continue to be kept as they are within that part of the United Kingdom now called Scotland; and that they shall so remain in all times coming.*' The Honours were lodged in the great oak chest, within the stone-vaulted Crown Room and all entrances were bricked up for additional

The Honours of Scotland.

Sir Walter Scott, 1771-1832.

security. In 1794 the Crown Room was opened up in the search for ancient documents, but the chest remained unopened and the entrance was once again walled up. This only strengthened the belief that the Honours had been secretly removed, possibly to London, as had the Stone of Destiny and *Mons Meg*.

The matter of the location of the Honours rested until it was raised by Walter Scott when dining with George, Prince Regent in April 1815. Subsequently, George granted a warrant for Scott and Officers of State to open the oak chest which was duly done on the 4th February 1818, revealing the Honours for the first time since 1707. The following year, paying visitors were admitted to inspect the Honours for themselves in the Crown Chamber where they have remained ever since, except when removed for presentation to reigning monarchs, the first time being during the state visit of George IV in 1822 and then not again until the visit of Queen Elizabeth II in 1953. Unlike the Stone of Destiny and *Mons Meg*, the Honours have never left Scotland.

George IV being presented with the keys to Holyrood Palace during his royal visit on 15th August 1822, the first by a reigning sovereign since 1651. Here he formally received the Crown, Sceptre, and Sword of State. In touching each emblem, he confirmed his kingship and connection with Scotland. Much of the royal visit was stage managed by Sir Walter Scott. When the king waved his hat to the crowd from the Half-Moon Battery at the castle, Scott took his opportunity to lament the absence of *Mons Meg*.

An Edwardian view of the Half-Moon Battery with *Mons Meg* in the distance.

The Galloway Legend

Andrew Symson, the Minister of Kirkinner, Wigtownshire was the first author to record the legendary origins of *Mons Meg*. In his account dating from 1684, he repeats what was by then already a well-established local verbal account that *'in the Isle of Threave, the great iron gun in the Castle of Edinburgh called Mons Meg was wrought and made.'* This traditional story has come to be widely known as the Galloway tradition or legend. Further elaboration was provided by Joseph Train (1779-1852), the Scottish antiquary and correspondent of Sir Walter Scott. Train worked as an Excise Officer in the Galloway region and collected southwestern tales and traditions and was at that time so widely respected as an antiquary that his version of the legend was soon accepted as authentic history.

The Galloway legend relates that in 1455, Threave Castle was the last stronghold of the powerful and rebellious Black Douglases, who had been pursued from one stronghold to another by a vengeful James II. However, at the time of the siege, James Douglas (1426-1488), 9th Earl of Douglas had

Trade card given away with chocolate bars by the German confectioner Stollwerk. This shows how bombards were placed on the ground for firing. Gunners were protected by means of palisades with opening shutters.

James II of Scotland with Edinburgh Castle in the background. The castle was, in 1440, the scene of the so-called Black-Dinner after which the 16 year old William Douglas, 6th Earl of Douglas, and his younger brother were beheaded. James. then ten years old unwillingly witnessed the whole murderous affair.

already taken refuge at the English Court, but the young king was so committed to destroying the Douglas rebels that he came in person to supervise and inspire the siege of Threave Castle. The royal army encamped at the "Three Thorns of Carlinwark" (where Castle Douglas now stands) and soon began the siege of the nearby castle. To batter the walls, the king brought with him certain early artillery pieces, known as bombards which were laid on the ground and their fire directed towards the curtain walls which were equipped with smaller defensive guns fired through embrasures.

Train's account relates that the locals were drawn to the siege procedures as spectators including a local blacksmith called M'Kim, who noticed that the king's artillery train was making little impression on the castle walls. He therefore boldly sought an audience with the king and offered to examine the small bombards with a view to making a much bigger version. These early artillery pieces were constructed of iron bars held together using iron hoops shrunk around the bars to create the barrel. Suitable iron bars were not commonly available, but the legend says that the townsfolk of Kirkcudbright, tiring of the tyrannical Douglas family, gladly provided them. M'Kim was aided in his task by his sons and they were said to have worked day and night to create the monster cannon.

Francis Groses's drawing of Threave Castle, showing the solid tower house and so called 'artillery house' built in the 1450s prior to the seige. This is among the first purpose-built artillery defences in Britain, consisting of a curtain wall with three towers. The gun ports can be seen clearly in the curtain wall and the towers could accomodate guns on two levels. While the gunners were reloading the slits could be used by archers. If all of this was not sufficiently formidable, it should be noted that the whole fortification sits on an island in the River Dee and the approach was through a marsh.

Once finished, the bombard was dragged to a commanding position right in front of Threave Castle, since known as Knockcannon. The first discharge was reported to have shaken the walls of the castle and induced a state of sheer panic in the defenders. The second firing propelled a ball through the thick castle wall and allegedly carried away the right hand of the Countess of Douglas, who was raising a goblet of red wine to her lips. This gory incident was said to be decisive and the garrison promptly surrendered without seeking terms. The final element of the story records that the grateful king rewarded M'Kim with the forfeited Douglas estate of Mollance which he then took as his new surname. The story neatly continues with the king naming this new bombard after its creator and his wife (Meg) resulting in *Mollance Meg*, later corrupted to *Mons Meg*.

Exactly where the Galloway legend ends and Train's elaboration begins is not easily determined, but the *Statistical Account of Scotland* repeats three strands of Train's so-called evidence:

> Two of these balls only are said to have been discharged at the siege and of both a satisfactory account can be given. The first, that which shook the castle and spread dismay among the garrison was, towards the end of the last century (18th), picked out of the wall and delivered to Mr. Gordon of Greenlaw; and in the year 1841, when removing for the purpose of husbandry, a large accumulation of rubbish from the lower part of the castle, he came upon a draw-well, which was found to be lined with planks of black oak in a state of perfect preservation, Prosecuting the search which this discovery provoked, the labourers at length came to an immense round ball, which on examination was found to be a bullet in all respects the same as those belonging to *Mons Meg* and still retaining evident marks of having been discharged from a cannon. It lay in the direct line from Knockcannon to the breach in the wall; so there is every reason to believe that this was the identical missile that shattered the Lady of the tremendous Lord of Galloway.

Unfortunately but not surprisingly, these immense cannonballs cannot now be traced. Further, Train's account does not take into account the subsequent sieges of Threave Castle in 1545 and 1640 when the enclosing walls and tower were extensively damaged and demolished. The second piece of evidence offered in support of the traditional story is said to date from the early years of the nineteenth-century. The castle was being prepared for use as a prison for French prisoners of war, when 'A massive gold ring, inscribed – Margaret de Douglas, supposed to have been on the Fair Maid of Galloway's hand when it was blown away at the siege, was discovered by one of the workmen.' As with the cannonballs, the ring is also elusive although it was said to have been given to the then Sheriff of the County, one Archibald Gordon. This section of the legendary account is a sensational invention and although Margaret Douglas is an interesting historical figure there is no evidence that she played any part in the defence of Threave Castle. She originally married her cousin, William, 8th Earl of Douglas who was stabbed to death by James II and his courtiers in Stirling Castle in 1452 at an early stage of the feud. Shortly afterward, the widowed Margaret married her dead husband's brother, James Douglas, 9th Earl of Douglas who was committed to avenging "the murder" of his brother by the King. James Douglas and Margaret were divorced during his exile in England and she married for a third time to the Earl of Athol in 1459.

The final piece of evidence is a little more credible and is said to date from the construction of the Military Road (now the A75) at Carlingwark (modern Castle Douglas) when workers found '*at the very spot where Mons Meg is said to have been manufactured …. a huge mound which turned out to be a mass of cinders, such are generally left by a forge.*' The cinders appeared to have been used in the construction of the road, but it is known that smiths did accompany the early bombards employed by the Scottish kings to carry on repairs in the field. If there is any truth at all in the Galloway legend then it is that bombards were repaired near the royal camp. J. M. Sloan observed that '*the entire story is beset with doubts as jagged as the Three Thorns of Carlinwark,* 'adding that the entire story '*had been inspired by the incipient idealism of a people struggling to escape from the night of serfdom and feudal barbarities towards the breaking dawn of civilisation proper.*' It was an elaborate attempt to associate the area with a supergun that had become a symbol of Scottish military power. As will be demonstrated later, *Mons Meg* was not of Scottish manufacture and arrived in Scotland two days after the siege of Threave Castle was over. After a siege lasting three months, the defenders of the castle finally surrendered in response to bribes and offers of safe conduct. No great cannon or even the smaller ones known to be in James II's artillery train were equal to the job of reducing the defences of this small island fortress. The last word on the legend should go to J.M. Sloan, who wrote '*if the story of Brawny Kim and Mons Meg at the siege of Threave must be doubted, it deserved to be true.*'

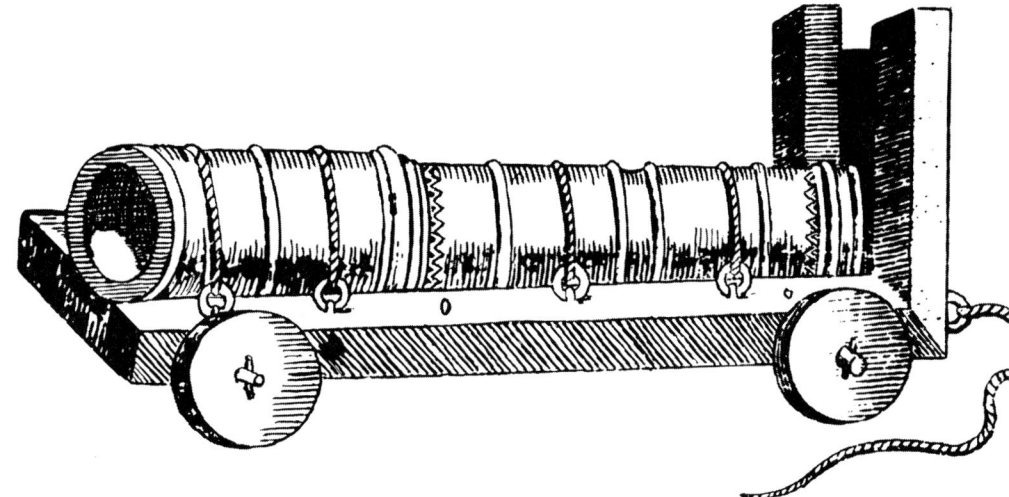

A seventeenth-century drawing of a bombard mounted on a simple carriage. The small wheels and low carriage indicate an adaption enabling use as an anti-personnel weapon in a fortification or possibly a naval setting.

This early twentieth-century postcard shows that bombards came in a great variety of shapes and sizes. Some of the later ones were manufactured with trunnions indicating that they were designed to sit on carriages or have a form of static support. (Musée de l'Armée, Hôtel National des Invalides, Paris)

Early Artillery in Scotland

In the earliest reference to the use of artillery in Scotland, the Scots were on "the receiving end" of this innovation in medieval warfare. The first Scottish War of Independence was punctuated by truces and in June 1327 one of these broke down and a Scottish force entered the English Marches. The fifteen-year-old Edward III led an army to repulse them which included a large number of Hainault mercenaries who brought with them early cannons although how and when they were used is not known. Edward's wife was Phillipa of Hainaut who came from the province whose capital was at *Mons* in what is now Belgium. The campaign went badly for the English and ended with a treaty in 1328 that recognised the independence of Scotland and Robert the Bruce as rightful king and cost the English treasury some £70,000 which included £41,000 for the services of the Hainaut mercenaries. Some four months afterwards another agreement was drawn up for the return of the Stone of Destiny, but the Scottish nation had to wait another 668 years for that to happen. John Barbour, the Scottish cleric and poet, described these early cannons as 'crakys of wer' which might be translated as loud noises of war. It is unlikely that these cannons played any significant part in the campaign but it does not take too much imagination to visualise the deep psychological effect that they exercised on friend and foe alike.

An illustration of one of these cannons dating from 1326 appears in a *Treatise on Kingship* commissioned by Isabella of France, Edward III's mother and is now to be found in the Library of Christ Church, Oxford. This shows a vase-shaped cannon with a long neck, standing on a trestle although it is certain that these weapons would have been bedded on the ground for firing. A mail-clad gunner is seen holding a red hot bar of iron to ignite the charge to propel a brass feathered iron bolt known as a *Garreau*. The shape and construction suggest a brass casting essentially the work of a bell founder.

A certain European, called Dietrich supplied '*ane instrument callit a gun*' for Edinburgh Castle in 1384 which is the first unquestionable use of firearms by the Scots. Doubtless, this was a defensive weapon for the castle and would have most probably been deployed around a gateway, projecting from a specially built gun slit. The English were already using guns in this way, as two years earlier the Prior of Drax had been commissioned to supply 'artellery' and gunpowder for the key English border fortresses at Roxburgh and Berwick. Earlier in 1338 at the start of the Hundred Years War, an English ship called *Christopher* was armed with three cannons and one handgun at the battle of Arnemuiden, the first recorded use of cannon in European naval warfare.

The first Scottish king to realise the usefulness of siege cannon was James I, who spent eighteen years as a prisoner of the English and accompanied Henry V on his second French campaign. Henry himself had

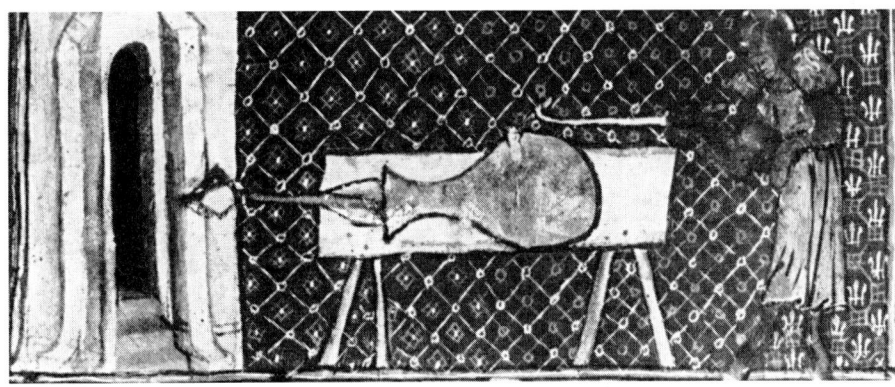

The earliest known illustration of a cannon from the manuscript treatise by Walter de Milemete. This type of cannon was used in Scotland by Hainaut mercenaries who formed part of the English force in 1327.

Above: James I of Scotland (1394-1437), the first Scottish monarch to buy artillery on a large scale.   *drawing by A.J.P. Butters*

Right: Henry V of England who employed early cannons in his French campaigns and was mentor to James I on matters of warfare.

Les Michelette two bombards abandoned by the English after the unsuccessful seige of Mont Saint Michel in 1434.

come to appreciate the potential of an artillery train during his Welsh campaigns and bombards were used to great effect on his first French campaign of 1415. Shakespeare made us familiar with the story of the Dauphin's insulting present of a barrel of tennis balls and of Henry's threat to reply with cannonballs. The famous breach at Harfleur was created by bombards called *Messager*, *The King's Daughter* and *London*. James witnessed the destructive power of the English artillery at the sieges of the towns of Beaugency, Dreux and Meaux. Two English bombards abandoned by Lord Thomas de Scales after the unsuccessful siege of Mont – Saint – Michel in 1434 are still displayed there. Now known as *Les Michelettes* these trophies are fine examples of the bombards used during this period.

James had played an active part in the war and was joint commander of the English force at Dreux even though a Scottish contingent was part of the French army. He witnessed the use of English bombards at the siege of Le Mans:

> the Englishmen approached at nighe to the walles as they might without their losse and detriment, and shot against their walles great stones out of great gonnes … the strokes of whereof so shaked, crushed and rived the walles that within a few days the citie was desployled of all her toures and outward defences.'

On his return to Scotland, James took possession of a rebellious kingdom and his priority was to restore royal power which meant the use of military force and the subjugation of his troublesome nobles. But he was also preparing for the inevitable renewal of hostilities with England. Artillery was essential and James was intent on buying the best which meant ordering in Flanders, then the European centre for the manufacture of ordnance. Here he purchased in 1430, a one and a half-ton brass bombard aptly called *The Lion* which was the most powerful weapon in the royal arsenal at that time. (For comparison *Mons Meg* weighs in at over six tons.) As a commissioned piece this bombard came with a proud inscription which translates from Latin as: '*to the illustrious James King of Scots, worthy prince, great king – only I reduce castles by basting – I was made for him, consequently I am referred to as The Lion.*'

During the early 1430s, the English were fighting desperately to hold the territories won in France by Henry V which allowed James to plan for and execute a more aggressive policy along his southern border. This included the commissioning of new ordnance and ammunition from Flanders and James paid a Nicholas Plummer some £590 in 1436 for making bombs and other military engines. James laid siege to Roxburgh Castle in the same year, setting his train of bombards the task of battering this key fortress into submission. These bombards were manned by German gunners under the command of Johannes Paule – 'Master of the King's Engines.' Sir Ralph Gray, the castle's governor was at the point of agreeing on terms to yield up the castle when James's queen arrived with news that caused him to lift the siege. She told him of a conspiracy ' which, if not speedily prevented, should endanger his Estate, person and Race.' William Drummond, the Scottish poet and historian, relates how the king '*found his imagination wounded upon this point (and) after many doubtful resolutions and conflicts in his thoughts, raiseth the siege, disbandeth the army and accompanied with some chosen bands of his most assured friends (returned) back to provide for his own safety.*' Accounts confirm that he left parts of his artillery train behind. To some Scots, this was a stain on the King's courage and honour but it should be remembered that James had first-hand knowledge of the plot in 1415 to assassinate Henry V as he assembled his army at Southampton. James was also keenly aware of the resentment among many of the Scottish nobility of his reassertion of royal power but the conspirators were only revealed when they succeeded in stabbing him to death in February 1437.

An illustration from *c.* 1470, showing a lifting frame and the heavy tackle needed to dismount a bombard for firing. A firing position would also have to be prepared. Clearly, this was a labour intensive activity hence the large numbers of pioneers that accompanied bombards. Leonardo Da Vinci envisaged a scene inside an arsenal with storage arrangements for bombards, about 1485-90.

Facing page: A German illustration from *c.* 1450 showing an attack on a castle. The illustration of the bombard is misleading as they would not have fired from their carriages at this time. Crossbowmen fire incendiary bolts. Three handguns (as would have been used at Threave in 1455) provide defenders with a means to hold off the attacking force.

"Fiery Face" and his Passion

The first King James had the spirit of a warrior king and a good technical knowledge of warfare, despite the disastrous siege of Roxburgh Castle. James II (nicknamed "Fiery face" because of a conspicuous vermilion birthmark) was only six at the time of his father's murder. It was to be a full fifteen years before he engaged in any significant military operations during the intermittent civil war which raged in Scotland from 1452 until 1455. The young king inherited his father's interest in artillery and when he began his military operations against the rebel lords in 1453, he used *'the great bombard'* (possibly *The Lion*) with a Sow to reduce the tower-house known as Hatton Castle, at Ratho 10 miles to the west of Edinburgh. William de Lauder supported the Douglas rebels against James which was to cost him his home and life. The Sow was a moveable protected shelter often equipped with a battering ram that allowed attackers to dismantle or undermine walls. Essentially Hatton Tower with its ten-foot thick walls fell to the twin effects of bombard fire and mining. An entry in the Accounts of the King's Chamberlain, dated 2nd July 1453, records payments made *'for four carts prepared for the carriage of the Great Bombard ….. for the siege of the House of Hatton.'*

A year later James turned his attention to Abercorn Castle, a Douglas stronghold 13 miles west of Edinburgh on the south bank of the Forth. An initial siege ended when the king withdrew his troops to face a Douglas threat elsewhere, but in 1455 a second investment began in earnest with his giant bombard and more traditional engines of war. One of the king's gunners, Allan Pantour, described as *'the most ingenious man in Scotland'* was killed by the explosion of one of the king's bombards a portent of a future event at Roxburgh Castle. The *Auchenleck Chronicle* relates in its archaic language how King James *'remanit at the sege, and gart strek doun with the great gun, the quhilk a Francheman schot richt wele, and falyeit na shot within a faldome (fathom) quhar it was chargit him to hit.'* After four weeks, the royal army stormed a breach and took the castle but the comment on the accuracy of the bombards is interesting *'within a (Scottish) faldome'* relates in more modern measurement to about 2 metres. The castle was subsequently dismantled and over the centuries used as a quarry for building stone and now exists as a mound in the grounds of Hopetoun House having been rediscovered through excavation in 1963.

These two siege operations were all within a 12-mile radius of Linlithgow Palace where the 'Magna Bombarda' and presumably other artillery pieces were kept. The expedition to reduce Threave Castle was an altogether greater logistical challenge involving transporting bombards a distance of around 100 miles on what was at best rough tracks. The route taken can be deduced from a broken wheel which had to be fixed at 'Crawfurdmuir', the then wild area centred on the modern village of Crawfordjohn in South Lanarkshire. The reference appears in the *Exchequer Roll*, which gives accounts for transport to and from Threave as well as some operational detail of the mounting of the 'Great Bombard' once there:

£10 – 17 Shillings Wheat to the Chancellor, William, Earl of Orkney coming with the Great Bombard

£5 To John Were (of Linlithgow) for carriage of the Bombard

£12 – 6 Shillings To Andrew Lisouris and John Were for renewing the wheel of the gun-cart broken In Crawfurdmure.

£22-5 Shillings for cutting timber and closing Bombard with it, and for iron, gunstanes, coals, spades and trowels.

13 Shillings 4 pence for a pair of wheels to a cart for bringing back bombards.

These accounts confirm that the bombards were carried in ordinary carts and placed on the ground for firing enclosed by a wooden protective frame to preserve the gunners from hostile fire.

The Galloway legend maintains that Threave Castle surrendered due to the effects of the artillery bombardment, but the Accounts of the King's Chamberlain include a payment made to the Steward of Threave Castle and other persons present in the castle at the time of the surrender. Two years after the surrender, payment was made for repairs to the 'artillery house' at Threave making good the damage inflicted by the royal bombardment. The royal bombards also suffered wear and tear during the siege of Threave and in 1457 John of Dunbar received £6 for repairing the great bombard at the Linlithgow arsenal. As will become clear shortly, this great bombard (*Magna Bombard*) was not *Mons Meg*. The question may also be asked whether it was *The Lion*, purchased by James I, and was this the same *Lion* that was said to have exploded during the siege of Roxburgh in 1460. This whole confusion arises because when a bombard was lost or became useless, its name was transferred to another piece of artillery, in much the same way that the Royal Navy has perpetuated the names of famous ships. A definite exception to this practice is provided by *Mons Meg*.

James II of Scotland, who had what was to prove a fatal passion for artillery. This contemporary illustration by Jörg von Ehingen shows the huge birthmark which earned the King the nickname of 'Fiery Face'.

The Duke's Gift

The short campaigns against James's nobles were set against the background of the emerging civil war in England that later became known as the War of the Roses after the publication of Sir Walter Scott's *Anne of Geierstein* in 1829. This conflict lasted with sporadic episodes of violence from 1455 until 1487, although there was related fighting before and after this period between the Lancastrian and Yorkist factions. James quickly recognised that this provided an opportunity to settle some of the old grievances with England. In August 1456, he made his 'first voyage in England', travelling twenty miles into Northumberland and destroying seventeen defensive towers although there is no record that he used any artillery on this expedition. It is known that John of Dunbar and Andrew Crauford were sent to Flanders to buy gunpowder and 8,800 pounds of iron for the manufacture and repair of Scottish ordnance. Flanders was then part of the Duchy of Burgundy and both were important trading centres for Scottish merchants. Burgundy was also officially at war with England since joining the French alliance in 1435. Both considerations had been significant when James chose as his bride, the niece of the Duke Philip of Burgundy, Mary of Guelders whom he married at Holyrood Abbey in July 1449. Neil Oliver has described the Duke as 'an international arms dealer' and observed that the marriage secured James 'his place at the top table of European power.' The Duke was the chief guest at the wedding and he had four weeks earlier received reports of the tests on a gigantic, newly completed bombard outside the city walls of Mons.

Duke Philip had ordered this bombard from Jehan Cambier, the foremost dealer in all kinds of armaments in the Low Countries and a Freeman of the City of Mons. Despite the alluring tale of M'Kim and his herculean efforts, the making of a large bombard was no simple matter and needed the facilities of a well-equipped forge like those to be found in abundance around Mons. To form the shape of the barrel a wooden core was employed on which long bars of iron were arranged before heated iron hoops were placed over them, cooled and so shrunk to bind the bars tightly together. This method of construction borrowed much from the art of the cooper and it is easy to see why the result was called a gun barrel.

Mons Meg was X-rayed by the Royal Armaments Research and Development Establishment in 1985 as part of a programme to finally determine how the bombard had been constructed. This examination

Philip Duke of Burgundy (1396-1467), uncle (by marriage) to King James II, who commissioned *Mons Meg* in 1449 and presented it to the King in 1457.

confirmed that the barrel was made of bars of iron and that these had not been welded together but simply butted up against one another. The powder chamber is made up of smaller pieces of iron hammer-welded together to form a single solid wrought-iron forging, the two parts of the bombard having been permanently fixed together with a tongue and groove arrangement pinched in by one of the iron hoops as it cooled. The powder chamber or breech obviously contained the charge of powder and so was generally cast or forged in one piece for additional strength. This method of construction may well have created a weak spot such that when *Mons Meg* was charged in 1680 with the more powerful gunpowder of that period it caused the then aged barrel to split.

There was another type of contemporary bombard made in two parts, the first of these being a gun called *Bourgogne*, manufactured for Duke Philip in 1436. The breech was made with a screw thread and square holes on the outside surface so that handspikes might be inserted to screw or unscrew the two-component parts as required. When *Bourgogne* was transported, each section was transported separately on two waggons, each drawn by 48 horses.

Mons Meg has a big sister in the shape of another great bombard preserved in Ghent, Belgium known as *Dulle Griet*, which translates as *Mad* or *Raging Meg* named after a figure from Flemish folklore. This is another medieval supergun founded in Mons made up of 32 longitudinal iron bars enclosed by 61 iron rings. It is known to have seen active service in 1452, when it was used by the citizens of Ghent in the siege of Oudenaarde. The relationship between these two guns was established by W. H. Finlayson, who, not content with the mere similarity of appearance, compared the dimensions of the two great bombards. A modified table of his findings appears below:

As Mr Finlayson observed 'a draughtsman would say that the two came off the same drawing board' and doubtless both were the work of Jehan Cambier or one of his subcontractors. The relative proportions show that these medieval two superguns were constructed in the ratio 4:5. Recent research by Robert D. Smith and Ruth Rhynas Brown has highlighted a third bombard from 'the same family' which might be described as a younger sister to *Mons Meg*. This is the so-called *Basel Bombard*, captured by the Swiss Confederates from Duke Charles the Bold of Burgundy at the Battle of Murten in 1476. This preserved bombard has similar proportions to *Mons Meg* and *Dulle Griet*. It is half the size of *Dulle Griet* and two-thirds the size of *Mons Meg*.

Mons was an expensive weapon at a final cost of £1,536 – 2 Shillings. The Duke was in no hurry to pay the bill or collect his new supergun and it was not delivered to him at Lille until May 1453 and then Cambier had to wait a further year for payment. During the first eight years of her existence, *Mons* never fired a shot in anger as the Duke had no immediate need for her services. Sometime before May 1457, he decided to present the bombard to his niece's husband, James II along with another smaller bombard and 3,000 lbs of gunpowder. At the port of Sluys, bombards, powder, and shot were loaded on a vessel bound for the port of Arnemuiden, then an important trading centre with Scotland. Two local shipmasters called Len Willemzone and Pietre Adrian contracted to provide two vessels for the shipment of the ordnance to Scotland. They left the port of Vere in Zealand in May 1457 with 50 mercenaries protecting against pirates and English naval vessels. The whole operation went smoothly and the landing was made at Leith before some sort of procession to Edinburgh Castle. The overall cost to the Duke of this lethal package was no less than £5,000.

Dimensions	(a) *Mons Meg*	(b) *Mad Meg*	Ratio (a to b)
Bore of Barrel	0.508 m	0.640 m	0.79
Bore of Chamber	0.216 m	0.260 m	0.83
Length of Barrel	2.807 m	3.390 m	0.83
Length of Chamber	1.130 m	1.375 m	0.82
Overall Length	4.064 m	5.025 m	0.81
Average			0.82

Dulle Griet or *Mad Meg*, 'sister to *Mons Meg*, preserved in Ghent.

An Edwardian postcard of *Mons Meg* at Edinburgh Castle.

The last siege of Roxburgh

There is no record of James's reaction to the Duke's gift, but considering his fascination with cannons he must have been overjoyed as he now had a weapon in *Mons* that was equal to the task of reducing all English fortifications on his southern border. James had launched an attack on the fortified town of Berwick in February 1457. This may account for the despatch of *Mons* from Flanders although a truce had been negotiated before she arrived in Scotland. During the period that the peace lasted, James engaged in negotiations with the English factions, *'plotting with Lancaster against York, while counterplotting with York against Lancaster,'* but all the time seeking to regain territory lost to the English. Finally, in July 1460 he gathered a large army and laid siege to Roxburgh Castle, the strongest and best defended of all the border fortresses. The castle was maintained at a cost of £1,000 a year in peacetime and £2,000 in time of war, so it was always well-maintained and provisioned. Roxburgh also represented a personal challenge to James as it was at Roxburgh that his father stumbled at the watershed of his reign in 1436. The timing of the Scottish attack was perfect as in England, the pace of the civil war had intensified with a series of costly battles beginning with that at Blore Heath in September 1459 and culminating with the bloodbath at Towton in March 1461.

The seventeenth-century Scottish historian, William Drummond provides a very complete account of the siege of Roxburgh. He reports the initial moves and the king's strategy shaped by the sure knowledge that no English relief force was to be expected. Drummond recorded that: *'The King's Army being gathered ….. passed the Tweed, invadeth the Town of Roxburgh which with little trouble is taken and equalled with the ground; the Castle, a strong fortress is besieged.'* Adding *'whilst the King here passeth the time, inviting it more by courtesies and blandishment than Ammunition and Warlike Engines to be rendered to him.'* James was preserving his force and military resources, perhaps influenced by his experiences at Threave. Meanwhile, Commissioners were received with a message from the Duke of York, *'requiring him to leave his siege, and contain himself within his own kingdom, unless he would run the hazard to engage himself in war against the*

The ruins of Roxburgh Castle seen from the River Teviot.

A Victorian illustrator's interpretation of the death of King James II at Roxburgh, 3rd August 1460.

whole body of the Kingdom of England.' These were hollow threats considering the situation south of the border, but this letter caused James to pursue a more active approach to the siege. Drummond reports the King as saying ' before their Embassie came I had resolved to take in and throw down this castle builded upon my bounds, and being by no benefit obliged to any of your Factions will not forwards leave off what I am about by arms to perform.' Immediately, he deployed his Battery (of bombards) against the castle *'which courageously defended itself, and holding beyond expectation bred an opinion that famine would be the only Engine to make it render.'*

Further reinforcements of men and materials of war continued to pour into the royal camp, including former rebels of the Douglas faction who brought with them *'a great company of (the) Irish (who) came to the Camp, men onely fit for tumultuous fights and spoil.'* In a move to mark the arrival of the Earls of Huntley and Angus on 3rd August 1460, an inspection of the siege works and bombards was organised and as Drummond tells us the King ordered *'a pale (in this context a fenced battery) of Ordinance'* to be discharged, one 'overcharged piece' which wounded the king in the thigh striking him 'immediately dead and the Earl of Angus forebruised.'

Francis Grose, the eighteenth-century antiquarian, military historian, and acquaintance of Robbie Burns, mentions that the bombard known as *The Lion* was at Roxburgh in 1460, but offers no authority for his statement. This bombard may have been abandoned to the English at Roxburgh after the 1436 siege and for this reason alone the name may never have been used again, but if it was *The Lion* which burst at Roxburgh in 1460 then it is obvious why the name was considered to be too tragically emotive to have ever been used again. Grose also stated that *'Mons Meg, cast at Mons in Flanders; was burst at the siege of Roxburgh and the piece was never used afterwards,'* an incorrect and potentially confusing statement. There are no contemporary records of *Mons* being at the second siege of Roxburgh, although it seems inconceivable that James would have missed this opportunity to deploy his most powerful weapon of war. Another possibility is that *Mons* was on the way to Roxburgh, to reinforce the smaller bombards that would have been more easily transported in carts. What is evident from all accounts is that James deployed numerous bombards against the walls of Roxburgh, his close supervision of which proved fatal.

At this point, the siege could have failed, but the widowed Queen Mary of Guelders rose to the occasion and came to the besieging army with her

Ruins of the entrance to Roxburgh Castle.

son Prince James, then only eight years of age. She revealed to the whole army the loss of the King with '*more than masculine courage caused new and desperate assaults to the Castle.*' The rest of Drummond's account reveals the methods being used by the attackers and the damage already caused – '*many turrets being shaken, some Gates broken, parcels of walls beaten down and mines ready in diverse quarters to spring*'. Ignorant of the death of the Scottish king but well aware that no relieving army would come soon, the garrison sought terms and surrendered. The Scots wasted no time in ensuring '*that it should not be a residence of aggression in following times is demolished and equalled with the ground.*' An old prophecy had foretold that a dead man would win Roxburgh Castle but it would be grossly unfair not to recognise the critical part played by Queen Mary of Guelders who plucked victory from potential defeat. Her victory also signified the end of the English occupation of Teviotdale and the permanent removal of the English military base at Roxburgh which in its day had been more important than that at Berwick. Only briefly, during the sixteenth-century conflict known in Scotland as the Eight or Nine Year War (but more commonly known as the Rough Wooing thanks to Sir Walter Scott) was the site at Roxburgh temporarily occupied by an English force.

The death of James had the potential to thrust Scotland into a state of chaos where factions would compete for control of the young king and the power to rule Scotland through him. Queen Mary again took the initiative and became Regent and the young king was crowned at Kelso Abbey on 10th August, a mere four miles from the site of his father's death at Roxburgh. Queen Mary continued effectively to rule Scotland until she died in 1463 and commenced the building of Ravenscraig Castle (in Kirkcaldy) in memory of her husband and as a dower house. Ravenscraig was originally conceived by James II as a fortification to withstand the work of bombards and deploy artillery in a defensive role. James III proved to be an unpopular and ineffective monarch due to an inability to deliver fair justice, his preference to seek alliances with England, and the breakdown of his relationships with virtually everyone in his family. However, he was not the pusillanimous character that some portrayed him to be and he shared his father's and grandfather's interest in artillery. He witnessed the final stage of the siege of Roxburgh and he rode with 'the great ordnance' when a Scottish army set out in 1463 intending to take Norham Castle. He also sponsored the first Scottish attempts to make the more efficient cast bronze artillery pieces in 1474, long before the first English trial castings of such cannon in 1521. James also met a violent death, in common with his immediate ancestors, when he was killed fighting the army raised by disaffected nobles at the Battle of Sauchieburn (near Stirling) on 11th June 1488. No specific references have survived of any active service by *Mons* during the reign of James III, although it is likely she was part of the ordnance train which set out from Edinburgh on the abortive strike towards Norham in 1463.

Surviving section of wall at Roxburgh Castle.

Serving James IV

During the second year of the reign of the teenage James IV, *Mons* was part of the siege train sent westwards to reduce the rebel strongholds, namely the castles at Duchal, Crookston, and Dumbarton. This is the first occasion that *Mons* is mentioned by name in any official royal records and the entry has an air of rejoicing: '*10th July (1489) gevin the gunnaris to drink – silver quhen (when) thai cartit Monss, be the Kings commande – Xviij s.*' Less than a month later, a further entry suggests that the siege train was on its return journey to Linlithgow: 4th August (1489) '*to Barcar and ane odir gunnar, to pass furth of Lythqow to Kirkyntloycht (Kirkintilloch) to help hame with the gunnis.*'

When James IV proposed to invade England as part of his package of support for the pretender to the English throne, Perkin Walbeck in 1497, *Mons* was made ready for action. Again the intended target was Norham Castle which guarded one of the fords on the River Tweed. A schedule of preparations is recorded in the *Treasurer's Accounts*. *Mons* and other artillery pieces were readied for war. In April, new wheels were made for the carts intended to carry *Mons* and other bombards and she was tested in Edinburgh Castle '*the casting (firing) of Mons geven by the Kingis command to the gunnaris.*' This brings to mind a future firing of *Mons* at Edinburgh in 1558 when a ball was projected from the castle walls. On the 21st of July, *Mons* left the castle for Norham accompanied by 100 pioneers to facilitate her passage and erect a protected emplacement once at Norham. It was a gay affair as musicians were engaged to '*playit before Mons,*' but such festivities were short-lived as her gun cart collapsed near St. Leonard's in the south of the city. It appears that the cart was given new wheels, but the vehicle itself was in a fairly decrepit state.

Little time was lost in making a new carriage or 'cradil as *Mons* lay by the side of the road. Timber was sent to St. Leonard's and seven joiners toiled for two and a half days on the work. Thirteen stone of iron was also used '*to make grath (bindings) to Mons new cradill*', which suggests that this was no cart but a carriage perhaps not unlike the one that *Mons Meg* rests on today. Tar and tallow were also supplied for mixing as a wheel lubricant, as was 'ane kinkin' (a keg) for storing this mixture. Another item of expenditure was for '200 spike nails to turse (pack) with *Mons*,' intended for use in constructing the protective wooden frame that would have been erected to protect the gunners from defensive fire once in position at Norham. Four great ropes were also supplied and *Mons* was drawn by carriage horses or oxen, whilst the 100 pioneers would have made improvised improvements to the rough roads over which she was hauled. John Mawer, the master joiner with two other joiners and two smiths made up the team which serviced *Mons*. A coarse cloth covered the bombard in transit and this had been painted with some sort of device by Sir Thomas Galbraith.

It may well have borne the arms of King James and the device indicates the esteem in which *Mons* was held – a royal supergun.

The Scottish army assembled at Upsettlington on 5th August 1497 and then crossed the River Tweed to lay siege to Norham. James IV is said to have played cards with Don Pedro de Ayala, the Spanish Ambassador to

James IV of Scotland, another keen exponent of artillery warfare.

Norham Castle commanded an important crossing point on the River Tweed. Despite its situation and heavy construction the keep was shown to be susceptible to artillery fire in the seige of 1513.

the Scottish Court while *Mons* and the rest of the king's artillery battered away at the walls of the castle. At best the siege was a half-hearted affair during which the Scottish assaults were repulsed and with the understanding (according to Drummond) that the *'Earl of Surrey was advancing with greater forces,'* King James lifted the siege and returned to Edinburgh. Some of the large balls fired by *Mons* during the siege were still at Norham Castle during the 1980s, along with an assortment of smaller stone shot possibly part of the consignment of 'gun-stanes' sent 'to Norham fra Edinburgh' on 4th August 1497. During the withdrawal, some of *Mons*'s 'bullets' were dumped in the River Leet, at Swinton Mill, where they were discovered in 1865 and in 1887.

The royal siege train paused at Dalkeith and the workmen were charged 'to bring hame' *Mons* and the other artillery pieces. *Mons* appears to have been then left on the ground somewhere in Edinburgh Castle without any covering. Four years later, work was ordered to clean away *'the erd (earth) fra Mons Meg and to turn her and lay the twych (touch) hole up.'* At the same time, a wooden shed was built for *Mons* and two other cannons called *Talbart* (*Talbot*?) and *Messenger*. These cannons had suspiciously sounding English names, but no more is known of them.

A truce was agreed in 1497 and lasted until 1513 when Henry VIII invaded France and James IV crossed the border in support of his French allies. By 22nd August, James was again besieging Norham Castle but *Mons* was not part of his artillery train. Instead, James took with him his new bronze cannons, including the legendary Seven Sisters. These superior weapons smashed down the barbican and the defences of the outer ward within two days and gave possession of the outer bailey to the Scots. As there was no immediate prospect of relief the garrison surrendered in keeping with the military practice of the time. The age of the giant bombard was passing and the new cannons were both easier to manoeuvre and more deadly in their execution of siege work. The chivalrous King James was killed a week after Norham Castle fell during the fighting on Flodden Field, some say because of his unwillingness to monopolise on his superiority in artillery.

Several of King James's bronze cannons had been sourced in France and Holland, but others had been cast in Edinburgh Castle by French gunfounders. James had appointed a Scotsman, Robert Borthwick to be Master Caster of royal cannons, and another persistent legend was that *Mons Meg* was made within the castle by him and that her *'first shot was fired as a salute on the day King James V was born.'* The grain of truth in this story was that *Mons Meg*'s prime function in years to come was firing royal salutes.

A section from the stone carving at Edinburgh Castle showing *Mons Meg* as she appeared *c.* 1500.

Royal Salutes and Exile

Between 1526 and 1539 further payments were made for the upkeep of *Mons*, typical examples being: '*to xii pynouris (pioneers) for taking and putting of the grete gun (on) her qhelis (wheels) – 3 Shillings*' and '*for overlaying Monce in the Castell with red lead – 30 shillings.*' *Mons* was still considered useful and her size continued to have the power to terrify those enemies of the Scottish crown who had the misfortune to see the wrong end of her barrel. At this time redundant bombards were deployed in castles as anti-personnel weapons, that is to say they could be used to fire stones, nails and other assorted small projectiles at attackers.

Exactly when *Mons* was first used to fire a royal salute remains unknown, but the first recorded occasion was to mark the marriage of Mary, Queen of Scots to the Dauphin of France which took place on 24th April 1558. Mary was married at Notre Dame de Paris so she did not hear the salute which was fired some weeks later. The ledger account (dated 3rd July 1558) records a payment to some pioneers for their labour in the mounting of '*Monss furtht of her lair to be schote*', and for the finding and carrying of the ball from Weirdre Muir back to Edinburgh Castle. It must have landed somewhere in the area now occupied by the New Town but the retrieval of 'her bullet' shows fitting regard for economy among royal officials. The reference to 'her lair' also suggests that the great bombard was not normally mounted but kept in a shed-like structure mentioned elsewhere in the royal accounts. The birth of James VI (later James I of England) in Edinburgh Castle on 19th June 1566 brought about a repeat performance. Five hundred bonfires were lit to illuminate the city and the surrounding hills, while all the artillery in the castle was discharged in salute.

The years following Mary's abdication were troublesome ones for Scotland. Edinburgh Castle was taken and held by Mary's supporters under Sir John Kirkcaldy. Richard Bannatyne (Secretary to John Knox) recorded how they took *Mons* down into the city in an attempt to terrorise their enemies:

> 14th May 1571. This day the grit canon was brought doun out of the Castell to the Blackfriar yaird; at (evening) the canon was careit up agane, whither for feare of thame without or no, I can not tell. But on the morne, being 15th of Maii, scho (she) was brought doune agane by the procurement of the Lord Huntlie, who was soveriet for her. Scho was stelled with gabions in the said

Mons Meg was fired fully charged with a shot to mark the first marriage of Mary, Queen of Scots. At the time she was still in Paris.

Mons Meg in 1982, when she had been transferred to one of the castle vaults and painted in red lead. It is possible that this was the sort of 'cradill' (carriage) constructed in 1497 when her gun cart collapsed on the way south to Norham. A copy of the stone carving and the plates from the iron gun carriage are also to be seen.

yard, which cost two or three poore men their lyves for the drawing of her. Scho shot this day twunty-four schot, ten whairof I saw and hard, shot at Lawsonis hous within two houris speace and a half.

Bannatyne's account is extremely useful in allowing an estimate to be made of the rate of fire that could be achieved with *Mons*. In the confines of the city, aiming would have been fairly straightforward as it is most likely that she was fired from prepared ground, protected by gabions, rather than from a carriage. Nevertheless, the charge had to be rammed down the barrel and packed with straw before the ball was rolled inside the vast chamber. Accepting that *Mons* could be discharged ten times in two and a half hours, then it must have taken an average of fifteen minutes to reload

her. The image of such a massive piece of ordnance being deployed within the confines of the city of Edinburgh is astonishing. The castle was finally yielded up by the Queen's supporters in 1573 and in an inventory taken two years later, there is a reference to *'ane grit piece of yron, callit Mons.'*

The first visual representation of *Mons* dates from the early seventeenth-century and is in the form of a carving set into the vaulted passage of the present outer gateway of Edinburgh Castle. *Mons* has pride of place alongside more contemporary examples of artillery pieces. *Mons* is shown on a gun carriage with massive wooden sides, resting on four spoked and studded wheels. Set against her mouth is a gunner's quadrant, implying that there must have been some means of raising or depressing the barrel, possibly by employing wedges. She was most likely more permanently mounted on this sort of carriage during the late sixteenth-century when her role had essentially become more ceremonial, but always with the option of use as an anti-personnel weapon. The present carriage on which she rests was copied from this carving and dates from 1934. More about this later.

Mons appears to have been completely neglected following her final period of active service in the streets of Edinburgh. The departure of James I for London in 1603 must-have severely limited the scope for the firing of royal salutes. On the occasion of the visit of Charles I to Edinburgh in 1633, she was considered unfit to participate in the salute of welcome, although an attempt was made to clean out *'the twichhole of the iron piece that had been poisoned thir many years bygane.'* *Mons* remained in the castle during the Civil Wars, despite another fanciful claim that she was removed to Dunnottar Castle to protect the Honours of Scotland when they were concealed there. *Mons* was listed with the other ordnance given up to Oliver Cromwell when Edinburgh Castle submitted to him after the battle of Dunbar in 1650. This well-known record reads *'the great Iron Murderer called Muckle Meg'* and elsewhere *'Muckle Magg.'* Despite being two hundred years old and obsolete, the sheer size of the bombard never failed to impress.

The reference to her as 'a murderer' is often misunderstood. By the seventeenth-century, 'a murderer' was a type of mortar so the name does not refer to any butchery done by *Mons*.

The substitution of Muckle for Mons is interesting as muckle in the Scottish tongue means large, so the name used in this account simply meant Large Meg. Cromwell's soldiers were impressed by her size, but what is interesting is the sudden appearance of the additional name Meg and this appears to have been contributed by the English soldiers of the Parliamentary army. *Roaring Meg* was the name given to a cannon

Above: James VI of Scotland and I of England, 1566-1625. His birth was celebrated by the discharge of all the artillery in Edinburgh Castle. A year later he became the infant monarch of Scotland following his mother's abdication.   *collection of Richard Muir*

Right: Although obsolete, *Mons Meg* was included in the inventory of ordnance surrendered to Oliver Cromwell in 1650.   *drawing by O.J. Lead*

employed in Staffordshire during the earlier stages of the Civil War and there is a mortar of the same name preserved on Castle Green, Hereford. On the walls of Londonderry is a yet another seventeenth-century cannon called *Roaring Meg*. Sometimes this "affectionate naming" of artillery pieces took a grim slant as in the case of *Sweetlips* captured by the forces of Parliament at Newark in 1646.

During the Cromwellian occupation of Edinburgh Castle, *Mons* maintained her prestige but the new appendage of Meg stuck. The Restoration of Charles II in 1660 was celebrated by the people of Edinburgh with real joy. The Castle Governor ordered the firing of 'the great cannon called *Mounce Meg*', followed by 'all *the guns in Edinburgh Castle, Leith Citadel and the ships in the road*.' This resumption of her former glory was short-lived and a year later, John Ray saw '*lying in the yard, a great old iron gun, which they called Mounts Meg and some Meg of Berwick*.' Why there should be a sudden connection with Berwick is unclear, although part of the Elizabethan fortifications at Berwick is known as Meg's Mount.

The active ceremonial role played by *Mons Meg* came to an abrupt and unexpected end in 1680, when firing a salute to welcome James, Duke of York (brother to Charles II) to Edinburgh. On his first visit to the castle, *Mons Meg* thundered out a welcome but the barrel of the two hundred and thirty year old bombard split. Onlookers took this as a bad omen and considering James's subsequent reign as James II of England, they were proved correct in their judgement. The Scots blamed it on the English gunner, '*thinking he might of malice have done it purposely, they having in all England no cannon as big as she*.' The Edinburgh poet, Robert Fergusson (1750-1774) echoed this belief nearly a century later when he wrote:

> "Oh willawins! Mons Meg for you;
> 'Twas firin' crack'd thy muckle mou';
> What black mischanter gart ye spew
> Bith gut and ga'?
> I fear, they bang'd thy belly fu'
> Against the law."

The truth of the matter rests in the truth that seventeenth-century gunpowder was considerably more powerful than that used when *Mons Meg* was constructed. It seems far more likely that the gunner was acting out of ignorance rather than malice. After this misfortune, she was unceremoniously abandoned in the castle where she is shown on Captain

To mark the restoration of Charles II, *Mons Meg* was fired in salute followed by all the other guns of the castle and those at Leith Citadel.

John Slezer's drawing from the late 1680s. Slezer commanded the Scots Train of Artillery sometimes known as the Artillery Company of Edinburgh Castle, but he was also Chief Engineer for Scotland and 'Surveyor of His Majesties Stores and Magazines.' For some unknown reason, he greatly exaggerated the size of *Mons Meg* which is curious as he must have known her very well. After Slezer's term of office, she was thoroughly examined in 1734 when an official report suggested that *'in all probability, (she was) the biggest gun ever made in the world.'* The same report also gives much detail about the dimensions and performance of the bombard:

> The length of her chase is 9 feet 2 and a half inches, of the chamber 3 feet 8 and a half inches, total length 13 feet 4 inches; diameter of bore 1 foot 8 inches; the weight of her bullet in iron 1125 lbs., in stone 549 lbs; her whole weight 19,452 lbs., or 8 and a half tons. It took 105 lbs. of Powder to fill her chamber when rammed; her greatest range was at an elevation of 45 degrees was 1408 yards with an iron bullet, and 2867 yards with a stone one; to travel these distances an iron bullet took 16 seconds and a stone one 22.

Twenty years later, an order was issued to General Humphrey Bland, Governor of Edinburgh Castle, instructing him to return all 'unserviceable' cannons to the central depot in the Tower of London. Apparently, following his orders to the letter, he included *Mons Meg* in the scope of this order and her departure was noted in the *Scots Magazine* (1st April 1754): *'On the 19th, a great gun in Edinburgh castle, called Mons Meg, said to weigh about 5 tons, and to be near two feet the bore, which has been long unserviceable, was carried thence to Leith, in order to be transported to London.'* She did not go easily and loading her on the merchantman *Happy Janet*, caused damage to the hawser and the ship itself. Recent research has shown that her fame had already spread and the removal from Edinburgh may have been more calculated than the action of an overly zealous military official. *The Derby Mercury* (24th May 1754), reported the arrival of the *Happy Janet* in the Thames, adding that on board was *'the famous gun, brought from Edinburgh Castle, called Mons Meg, which for size is equalled by none, except one at Ghent in Flanders.'* Before she arrived in London, her uniqueness was already recognised and she began a planned new role as a tourist attraction for visitors to the Tower. Captain Francis Grose knew her well as an exhibit mounted awkwardly on a standard eighteenth-century field gun carriage. She was a Scottish national treasure in exile.

James VII of Scotland and II of England. When James, as Duke of York came to Edinburgh Castle in 1680 he was greeted by a general artillery salute. *Mons Meg* had been overcharged and this cracked her barrel ending her final active role.

Captain John Slezer's drawing made in the late 1680s showing *Mons Meg* abandoned inside Edinburgh Castle.
City of Edinburgh Council Edinburgh Libraries

Home Again

Although exiled a long way from Edinburgh, *Mons Meg* was not forgotten by the folk of her adopted homeland. Robert Fergusson (1750-1774), a poet of the Scottish Enlightenment and influence on Robbie Burns was only four years old when *Mons Meg* was taken to London, but this did not prevent him from eulogising on her prowess:

Right seenil am I gien to banning;
But, by my saul, ye was a cannon could hit a man, had he been stannin,
In shire o' Fife, Sax lang Scots miles ayont Clackmannan,
An' tak' his life.

Burns wrote an election ballad during his creative and fruitful time at Ellisland Farm which references the bombard:

O for a throat like huge Mons Meg
To muster o'er each ardent Whig
Beneath Drumlanrig's banners.

Other remembrances of her were less respectful, a ballad called the *Blythsome Bridal* written before 1706 contains the following lines:

And there will be fairteckl'd Hew and Bess with lily-white Leg,
That gade to the South for Breeding and bang'd up her Warme in Mons Meg.

Francis Grose knew of this ballad and it is tempting to imagine him hearing it from the person that he termed '*my ingenious friend, Mr. Robert Burns*' following their meeting at Friars Carse in 1789. Burns suggested that Grose include Alloway Church (where his father was buried) in his projected *Antiquities of Scotland*. Grose agreed provided that Burns wrote a witch story to accompany the drawing. In 1790, Burns sent Grose a prose tale followed up by a rhymed version "Tam o' Shanter.' The relationship seems to have cooled later when Burns wrote a poem on the subject of Grose's travels in Scotland warning his fellow Scots that there was '*a chiel (lad) amang ye takin' notes.*'

Even to this day, *Mons Meg* is officially and to some inexplicably, still part of the Armouries Collection of the Tower of London. During her exile

Robert Burns, (1759-1796), national poet of Scotland. He wrote about *Mons Meg* but never saw her.

Captain Francis Grose, FSA, who wrote *A Treatise on Ancient Armour and Weapons* (1786) and *The Antiquities of Scotland*. He meticulously examined *Mons Meg* during her sojourn at the Tower of London.

Grose's sectional drawing of *Mons Meg*, c. 1786. She is shown on eighteenth-century field carrige which made the journey north with her in 1829.

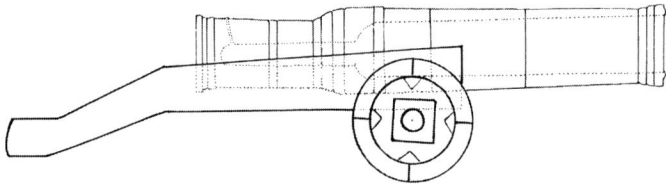

there, Scots were reported to have lamented that '*Scotland would never be Scotland again until Mons Meg came hame.*' The first move to bring her back to Edinburgh was made by Sir Walter Scott, who had been already steeped in the bombard's legendary origins by his friend and correspondent, Joseph Train. Scott's opportunity finally came when George IV agreed to visit Edinburgh in August 1822 and Scott already well known to the royal family seized the opportunity to stage a splendid pageant that was intended to allow the ancient identity of Scotland to be reborn. When George waved his hat to the crowd from the Half Moon Battery at Edinburgh Castle, Scott's biographer, Gibson Lockhart tells us that: '*when Scott next saw the King, after he had displayed his person on the chief bastion of the old fortress, he lamented the absence of Mons Meg on that occasion, in a language which His Majesty could not resist.*' Less than a month later, the Duke of Wellington, then Master General of the Ordnance wrote to Sir Robert Peel who was the intermediary in this matter: '*this gun is in the Tower and is one of the principal articles in Singleton's Show. But I shall have no objections to it being sent to Edinburgh Castle – you may tell Sir Walter Scott that it shall be sent to Edinburgh, but I must get the King's orders to remove it from the Tower.*'

At this stage, Scott's initiative lost momentum, possibly because of the Duke's absence abroad but also as Wellington may have preferred to move *Mons Meg* to the Royal Repository at Woolwich, which he had chosen as the most suitable place to display the cannons captured at Waterloo. The Duke saw the Tower of London as the place to store 'those arms and stores calculated for use in modern times' rather than as a museum of antique ordnance. Sir Walter was persistent and his diary for December 1825, records his resolution to visit London during the Spring and renew '*my negotiations with the Great Duke for the recovery of Mons Meg.*' Scott visited London in 1826 and dined with Wellington on more than one occasion, but his renewed requests did not immediately bring the desired result. Indeed it was a completely new initiative in 1828 which finally secured the return of *Mons Meg* to Edinburgh. These representations came from the Society of Antiquaries of Scotland who recruited the Duke of Gordon, Governor of Edinburgh Castle as their main agent. His formal request for the return of the bombard in April 1828 brought to light another reason why Scott's campaign failed. Apparently, in 1822 the inhabitants of Stirling had also expressed a desire to have *Mons Meg* for their own claiming that before it was in Edinburgh it had stood on their Castle Green. The only evidence to back up their opportunistic claim was the former existence in Stirling Castle of a battery called Meg's Mount. The matter was soon resolved in favour of the Edinburgh lobby and in June 1828, George IV gave his royal permission for *Mons Meg* to be removed from '*his Tower of London to his Castle of Edinburgh*' and placed at '*the disposal of the Antiquarian Society of Scotland*'.

Above: Sir Walter Scott, who campaigned for the return of *Mons Meg*.

Right: An early photograph of the Duke of Wellington as Master of the Ordnance and Constable of the Tower of London. Scott quickly realised how crucial Wellington would be in securing the return of *Mons Meg* to Edinburgh Castle.

A very full account of the return of *Mons Meg* to Edinburgh appeared in *The Scotsman* (11th March 1829):

> After an absence of nearly seventy-five years, this ponderous mass of antiquarian ordnance was on Monday replaced in our venerable fortress with all due honours. It was on the 24th April, 1754 that this piece of antiquity left the castle and was drawn down the Canongate and thence by Easter Road to Leith, where she was shipped on board the Happy Jane for the Tower of London, where she lay neglected till the Antiquarian Society interested itself with success to get her restored to her ancient domicile. Some time ago we mentioned her arrival and landing at Leith from the City of Edinburgh steam –packet, whose proprietors gave her gratuitous passage, since when she has lain in the Naval Yard.
>
> As it was generally known that Monday was the day fixed upon for her removal, a great concourse of spectators assembled. A troop of the Third Dragoon Guards, a party of the Royal Artillery, and a strong detachment of the Seventy-eighth Highlanders under the direction of Major Broke, Assistant Quartermaster General were in attendance to escort Meg to her old quarters, which at ten minutes past twelve (proceeded by some members of the Celtic Society) left the Naval Yard drawn by ten horses decked with ribbons and evergreens, the two leading horses being rode by two boys dressed in tartan and carrying broadswords. The line of march was that adopted on the landing of His Most Gracious Majesty [George IV], viz, by Leith Walk, York Place and St Andrew's Square, and thence by North Bridge to the Castle, where the Royal Standard was hoisted in honour of the occasion, the gates being closed and all other ceremonies being duly observed. At half-past one o'clock the advance guard gave notice of Meg's approach, when she was welcomed by the hearty cheers of a dense multitude of all classes, the band of the Dragoons playing Highland Laddie, which on her entering the gate was changed to 'God Save the King'. She was then drawn to the Argyll Battery and placed on a carriage prepared for her reception in front of the Main guard-house.
>
> There were a number of carriages on the hill, among which was that of the Countess of Hopetoun, the rest chiefly belonging to military persons. The day was fine and the bells of St. Giles lent their aid to enliven the gay scene, nor must we omit to state that a flag was displayed from the top of Marshall's Panorama, who also fired guns in honour of the occasion.

Sir Walter Scott and his family were in attendance and his diary entry for the 9th March reads as follows:

> went about one o'clock to the castle, where we saw the auld murderess *Mons Meg* brought in solemn procession to occupy her ancient place in the Argyll Battery. The day was cold but serene and I think the ladies must have been cold enough, not to mention the Celts (Celtic Society) who turned out upon the occasion under the leading of Cluny Macpherson, a fine spirited lad. *Mons Meg* is a monument to our pride and poverty. The size is enormous, but six smaller guns would have made at the same expense and done six times as much execution. [Scott's daughter] Anne had a narrow escape when some rockets were launched 'one of which lighted on her head and set her bonnet on fire. She neither ran nor screamed, but quietly permitted Charles Sharpe… to extinguish the fire.

Mons Meg was now sited on the Argyll or Six-Gun battery, facing northwards on the aged wooden artillery carriage shown in Grose's drawing of 1788. Even before her return, the Antiquaries of Scotland received an accurate, large drawing from Lieutenant Battersbee of the Royal Engineers with a view *'to construct a proper iron carriage for the gun.'* They also interested themselves in the *'seven stone shot ordered to be destroyed in 1754'*, that had survived *'located in various situations within the castle and the greater part of them in perfect state'* and had them *'piled on the battery next to Mons Meg.'* (*Caledonian Mercury*, 31st January and 18th April 1829.)

The *City of Edinburgh* steamship that brought *Mons Meg* 'hame' from London in 1829.

Admiring Queens

Despite the good intentions of the Antiquarian Society of Scotland, *Mons Meg* remained on the decrepit gun carriage until the night of Wednesday, 10th June 1835. A ridiculous article in the *Edinburgh Observer* set out to sensationalise what was a rather mundane event referring to spectres, fairies, and the 'ghost of *Mons Meg*'. What had happened was that the aged carriage finally gave way under the immense weight of the bombard. Not for the first time, '*Old Mons is prostrate on the fort from whence it formerly threw its stone-bullets to the sea-shore at Wardie Muir.*' The Officer commanding 'Artillery in North Britain' counselled that an iron carriage would be a better option than another wooden one. His advice was accepted by the Master-General of Ordnance and the wooden pattern needed for casting an iron carriage already existed. It had been originally ordered for the manufacture of an iron carriage for the 'Turkish Gun' captured in the forcing of the Dardanelles Straits in 1807. This huge cannon and carriage can still be seen on the grounds of Admiralty House, Devonport.

Queen Victoria and Prince Albert inspecting *Mons Meg*, 1842. Lieutenant-General Neil Douglas, Governor of Edinburgh Castle is shown recounting the bombard's history

The new iron carriage for *Mons Meg* was cast in London for £53 and brought to Edinburgh by sea. It weighed 3½ tons and provided the sturdy support that was previously lacking. A more customised commission, the carriage was decorated with a thistle; the star of the Order of the Thistle; a crown, and 'W[illiam]. the IV Rex.' Additionally, a few historical facts were recorded on small plates affixed to the carriage (at the request of the Society of Antiquaries of Scotland: 'At the siege of Norham Castle. A.D. 1497. Sent to the Tower of London A.D. 1754. Returned to Scotland by H.M. Geo. IV 1829.' *Mons Meg* was mounted on her new carriage sometime shortly after November 1836. She was also moved from the Argyll Battery to the Mortar Battery, standing before St. Margaret's Chapel with her barrel pointing towards Princes Street.

In September 1842, on her first visit to Scotland, the young Queen Victoria visited Edinburgh Castle. A contemporary account describes how the royal party '*first proceeded to the Mons Meg Battery with the approach to it covered with a crimson cloth.*' From here she was able to look out over Edinburgh and '*Her Majesty remained here about ten minutes admiring the fine prospect and minutely inspecting Mons.*' Then, the Queen accompanied by Prince Albert went to inspect the Honours of Scotland. The contemporary illustrations, one from the recently founded *Illustrated London News* and another, a contemporary print show Victoria arm in arm with Prince Albert, being told the history of *Mons Meg* by Lieutenant-General, Sir Neil Douglas, then Governor of the Castle. The iron and stone cannonballs for *Mons Meg* are also shown, along with an eighteenth-century mortar added to complete the martial scene.

Afterwards, *Mons Meg*, became a must-view part of the Edinburgh experience as tourists followed the example of their Queen in visiting Edinburgh and Scotland in general. In Australia, a cavern was named after her due to the resemblance of one of the limestone columns to the ancient bombard. Kirkcaldy in Fife called their fire engine after this revered artillery piece and in the early 1890s, there was a noted racehorse called *Mons Meg*.

In 1865 and 1887, stone cannonballs were found in the River Leet, near Swinton Mill in Berwickshire. They were of the correct bore to make an immediate link with *Mons Meg* and the ill-fated Norham Castle campaign of 1497. Some were thought to have been dumped there but others were thought to have been in a waggon that got stuck whilst fording the river as the Scottish army made a hasty withdrawal in the face of a larger English force.

Mons Meg just prior to being mounted on the 'new' iron carriage, 1836.

In July 1934 King George V and Queen Mary made their seventh State Visit to Edinburgh and this prompted the revival of an earlier idea to sit *Mons Meg* on a more fitting carriage. Most medieval bombards were transported on heavy carts but it seems that in 1497 a new purpose-built 'cradil' was constructed for the gun as it made its way southwards to Norham. The appearance of this cradle or carriage is known from the stone carving on the inner side of the gateway arch of Edinburgh Castle. The bombard was always intended to be fired from the ground, but the carriage would have provided a much easier way of transporting it. In the 1930s, it was felt that the iron carriage gave '*a false impression of what the famous gun looked like*' in the days of her active service and should be replaced. A design already existed and in May 1934, work was commissioned by engineer and businessman, Sir William Johnston Thomson, who served as Provost of Edinburgh from 1932 to 1935.

The Scottish oak for the transoms came from the Minto estate, the sides of the carriage from two ancient oaks grown at Swinton, five miles from Norham. The carriage was built by J. and T. Scott of Edinburgh at a total cost of £178, met entirely by Sir William. In a short ceremony before the royal visit, the carriage was officially handed over to the Office of Works by Sir William and Lady Thomson. On the arrival of the royal couple, it was the Queen who inspected *Mons Meg* mounted on the new but authentic carriage and she thanked the Provost and his Lady for their generosity. The fate of the now redundant iron carriage is uncertain but one source says that it was scrapped during the Second World War, something that might have befallen *Mons Meg* herself. In May 1942, H.M. Office of Works issued a statement assuring the public that the ancient bombard was safe and

A poor photograph, but one of historical importance. Queen Mary meets Sir William Thomson, who financed the replacement carriage, and his Lady.

would not be scrapped. The plates from the iron carriage were on display beside the gun during the 1980s when she resided briefly in the *Mons Meg* Vault and was preserved with 'reid leid' paint as she had been in the early sixteenth-century. That is how I first saw her in November 1982 in the company of the Muir family.

The seventeenth-century panel recording Edinburgh Castle's role as a principal arsenal. The representation of *Mons Meg* was used to design the replacement wooden carriage built in 1934.

A portion of the grand print made from a painting by David Octavius Hill, published in 1857. In addition to being a very accomplished artist, Hill is world-renowned for his pioneering photographic work.

Images and Souvenirs

It is striking how many Victorian paintings of Edinburgh from the castle manage to incorporate *Mons Meg* in the composition, plus a plethora of prints and postcards show it sitting on one or other of its modern carriages. Likewise, there has been a multitude of souvenirs featuring *Mons Meg* ranging in size from bracelet charms to scale models of the bombard itself in a range of materials. In this respect, it could be argued that it is better represented than either the Stone of Destiny or the Honours. 'The Great Gun' as it has so often been described has embedded itself in the imagination and memory of visitors for centuries and it is arguably the most famous cannon in the world.

The earliest image, drawn by Captain John Slezer around 1690, shows *Mons Meg* massively exaggerated in size lying abandoned inside the castle. She is being examined by an officer and a sergeant, perhaps in the spot where she had been dumped after the ill-fated salute which cracked her barrel some ten years earlier. A more scientific drawing dates from 1788 when she was at the Tower of London and mounted on a contemporary wooden carriage.

Nearly all the Victorian paintings and prints of the view towards Edinburgh from the castle feature *Mons Meg*. Many show her

A selection of *Mons Meg* souvenirs. A: A pewter model on her nineteenth-century carriage, overall length 18 cm. B: Early twentieth-century paperweight, length 10 cm. C: Mauchline ware money box – white wood with a printed transfer, made in Mauchline, Ayrshire – height 18.5 cm. D: Scale replica by Alexander Pollock of Edinburgh, overall length 13 cm. E: Early twentieth-century Goss type pottery piece. This one bears the arms of the City of Edinburgh but examples are known named for towns and cities across the United Kingdom, overall length 13 cm.

surrounded by soldiers from whatever regiment formed the garrison at the time. A particularly fine example is the print based on a painting by David Octavius Hill, now in Perth Museum and Art Gallery, published in August 1857. At the centre is an old soldier reading newspaper accounts of the actions in the Indian Mutiny, surrounded by serving Highland soldiers. Full of fascinating detail, Hill manages to incorporate a group of fishwives derived from one of his early photographs as were a number of the soldier figures. Sat astride, *Mons Meg* is a small girl steadied by her mother's hand.

American visitors have frequently taken *Mons Meg* to their hearts and in one instance this led to the bombard being featured in a 1924 advertisement for Reading (Pennsylvania) Wrought Iron Pipe. Claiming falsely that the bombard had been '*unprotected by grease or paint, she has braved all weathers for four hundred years and her surface is hardly pitted.*' The advertisement continues '*Remarkable? Not when we remember that she is made of wrought iron.*'

An American businessman, pictured with his wife and daughter beside *Mons Meg*, in the late 1920s. His son-in-law is the photographer represented by the hat on one of the bombard's 'bullets'.

The 1924 advertisement inspired by an American tourist's visit to Edinburgh Castle. *Mons Meg* became the subject of some inflated claims for the virtues of wrought iron made by the Reading Iron Company of Reading, Pensylvania.

LNER No. 2004 *Mons Meg*, built to haul Edinburgh to Aberdeen trains. This photograph was taken after the locomotive was rebuilt in November 1944.

In the early 1930s, *Mons Meg* featured in a fine work by Frank Newbould, one of the leading inter-war designers particularly known for his bright colourful travel posters. This one (*inside back cover*), produced for the London and North Eastern Railway Company (LNER), shows a fashionable couple enjoying the sights of Edinburgh, but especially *Mons Meg* mounted on her iron carriage.

The LNER also commissioned the 'P2' class of locomotives designed by Sir Nigel Gresley for working express trains between Edinburgh and Aberdeen. Six locomotives were built between 1934 and 1936, all given names from Scottish lore. In July 1936 the locomotive *Mons Meg* made its first appearance on the railway system and remained in service until 1961. In 1984 a new electric locomotive entered service which still proudly displays the name of this national treasure.

The rather splendid *The History of Scotland For Children* by Fiona Macdonald (2014), features *Mons Meg* on its cover, along with Kenny Dalglish, the Scottish Thistle, St Martin's Cross, Iona, and Bonnie Prince Charlie. Historic Scotland's *Edinburgh Castle for Kids*, (2019) features a page of instructions on how to fire *Mons Meg* describing her as '*the World's most famous medieval cannon.*' Both confirm the bombard's continuing popularity with the younger generation.

An animated scene around *Mons Meg* in 1929.

Stereoscopic card from 1861 of *Mons Meg*.

Further Reading

The following were the principal books used by the author during his research. None are available from Stenlake Publishing; please contact your local bookshop, reference library or search for them on the internet.

BOOKS

John Anderson, *A History of Edinburgh from the earliest period to the completion of the half-century (1850)*, (Edinburgh 1856.)
William Anderson, *The Scottish Nation*, (Edinburgh, 1865.)
George Bennett (Editor), *The Exchequer Rolls of Scotland*, (Edinburgh, 1878-1901.)
H.L. Blackmore, *The Armouries of the Tower of London*, (London, 1976.)
C.H. Hunter Blair and H.L. Honeyman, *Norham Castle* (London, 1976.)
Charles J. Burnett and Christopher Tabraham, *The Honours of Scotland*, (Edinburgh, 1993.)
David H. Caldwell, *Scottish Weapons and Fortifications, 1100-1800*, (Edinburgh, 1981.)
W. Y. Carman, *A History of Firearms from Earliest Times to 1914*, (London, 1955.)
Stewart Cruden, *The Scottish Castle*, (Edinburgh, 1960.)
Thomas Dickson, *Accounts of the Lord High Commissioner of Scotland*, (Edinburgh, 1877.)
William Drummond, *A History of Scotland from the year 1423 until the year 1542*, (London, 1655.)
Alexander Eddington, *Castles and Historic Homes of the Border*, (Edinburgh, 1946.)
James Faed and J.M. Sloan, *Galloway*, (London, 1908.)
Richard Fawcett, Iain McIvor and Nicholas Reynolds, *Edinburgh Castle*, (Edinburgh, 1980.)
Antonia Fraser, *Mary Queen of Scots*, (London, 1969.)
James Grant, *Memorials of the Castle of Edinburgh*, (Edinburgh, 1850.)
James Grant, *Old and New Edinburgh: Its History, Its People and Its Places*, (Edinburgh, 1880.)
W. Forbes Gray, *Edinburgh Castle*, (Edinburgh, 1948.)
Francis Grose, *Military Antiquities*, (London, 1788.)
Francis Grose, *The Antiquities of Scotland*, (London, 1789-1791.)
J. Haigh, *A Topographical and Historical Account of the Town of Kelso and the Castle of Roxburgh*, (Edinburgh, 1825.)
Lt. Colonel Hennebert, *L'Artillerie*, (Paris, 1887.)
O.F.G. Hogg, *English Artillery, 1326 – 1716*, (London, 1963.)
Peter Lead, *Mons Meg: A Royal Cannon*, (Holmes Chapel, 1984.)
R. Lindesay of Pitscottie, *The Historie and Chronicles of Scotland*, (Scottish Text Society, 1899-1911.)
John Gibson Lockhart, *Life of Scott*, (Edinburgh, 1838.)
R.L. Mackie, *King James IV of Scotland*, (Edinburgh, 1958.)
John Menzies, *Extracta Ex Chronicis Socie*, (Edinburgh, 1842.)
Ranald Nicholson, *Scotland: The Later Middle Ages*, (Edinburgh, 1974.)
Robert Norton, *The Gunner: The Making of Fireworks*, (London, 1628.)
Neil Oliver, *A History of Scotland*, (London, 2009.)
J.S. Richardson and Marguerite Wood, *Edinburgh Castle*, (Edinburgh, 1953.)
W. Douglas Simpson, *Threave Castle*, (Edinburgh, 1967.)
Robert D. Smith and Ruth Rhynas Brown, *Bombards: Mons Meg and her sisters*, (London, 1989.)
Christopher Sinclair-Stevenson, *Blood Royal*, (London, 1979.)
Andrew Symson, *Large Description of Galloway*, (Edinburgh, 1823.)
Anthony Tuck, *Border Warfare*, (London, 1979.)
Alexander Warrack, *Chambers Scots Dictionary*, (Edinburgh, 1982 reprint.)
Francis Watt, and the Reverend Andrew Carter, *Picturesque Scotland: Its Romantic and Historical Associations, Described in Lay, legend and Song*, (London, 1889.)
M.G. Williamson, *Edinburgh: Historical and Topographical Account of the City*, (London, 1906.)

ARTICLES

R. N. Appleby-Miller, 'Mons Meg: A 15th Century Bombard', *Transactions of the Dumfriesshire and Galloway Natural History and Antiquarian Society*, Vol. XXI (1939), pp. 360-369.
Claude Blair: 'A New Carriage for *Mons Meg*', *Journal of the Arms and Armour Society*, Vol. V, No.12, (December 1967), pp. 431-452.
R.C. Clephan, 'Early Ordnance in Europe', *Archaeologia Aeliana*, Vol. XXV (1904), pp 1-61.
W.H. Finlayson, '*Mons Meg*', *Scottish Historical Review*, Vol. XXVII (1948), pp. 124-6.
Claude Gaier, 'The origin of *Mons Meg*', *Journal of the Arms and Armour Society*, Vol. V, No.12 (December 1967), pp. 425-431.
W. Forbes Gray, '*Mons Meg*', *The Scotsman*, 9th March 1929. Sir James Balfour Paul, 'Ancient Artillery: with some notes on *Mons Meg*', *Proceedings of the Society of Antiquaries of Scotland*, Vol. L (1915-16), pp 191-201.